# Understanding the UK Mathematics Curriculum Pre-Higher Education

– a guide for academic members of staff –

**Understanding the UK Mathematics Curriculum Pre-Higher Education**

# Contents

**1. Background** ........................................................................................................................ 3
   1.1 The rationale for the document ............................................................................ 3
   1.2 Introduction and overview ...................................................................................... 3

**2. Setting the scene: pre-higher education qualifications and study** ............. 4
   2.1 Introduction to the main qualifications ................................................................ 4
   2.2 Brief historical review of major developments .................................................. 4
   2.3 Where and how will entrants have studied pre-higher education? .............. 5

**3. Specific UK qualifications and attributes of students who enter higher education with them** ........................................................................................................ 5
   3.1 General Certificate of Secondary Education ..................................................... 6
      3.1.1 Overview ........................................................................................................ 6
      3.1.2 Subject knowledge and skills ....................................................................... 6
      3.1.3 International General Certificate of Secondary Education ................. 7
   3.2 Advanced Subsidiary and Advanced Levels ........................................................ 7
      3.2.1 Overview ........................................................................................................ 7
      3.2.2 Subject knowledge and skills ....................................................................... 9
   3.3 Advanced Extension Award and Sixth Term Examination Paper ................... 9
   3.4 Free Standing Mathematics Qualifications ........................................................ 10
      3.4.1 AS Use of Mathematics ............................................................................. 11
   3.5 Diplomas ................................................................................................................... 11
   3.6 Other qualifications ................................................................................................ 11
   3.6.1 International Baccalaureate .............................................................................. 11
   3.6.2 Pre-U ..................................................................................................................... 12
      3.6.3 Access courses ............................................................................................ 12
      3.6.4 Foundation courses .................................................................................... 12
   3.7 Wales, Scotland and Northern Ireland .............................................................. 12

**4. Useful sources of information (to keep up-to-date with pre-higher education developments)** ............................................................................................... 13
   4.1 References made in this guide ............................................................................. 14
   4.2 Additional references ............................................................................................ 14
      4.2.1 Documents/information ............................................................................. 14
      4.2.2 Organisations ............................................................................................... 15

**5. Appendices** ..................................................................................................................... 17
   5.1 Acronyms used in this guide (including appendices) ...................................... 17
   5.2 A Level Mathematics Numbers 1989 – 2009 (Source JCQ) ........................ 18
   5.3 Overview of content in mathematics A Levels ............................................... 19
      5.3.1 What mathematics do students study in A level Mathematics courses? ... 19
   5.4 Important dates for Mathematics (authored by Roger Porkess) ................ 23

# 1. Background

## 1.1 The rationale for the document

In order to study a wide range of undergraduate programmes (including those in the Biological Sciences, Chemistry, Computer Science, Engineering, Materials Science, Mathematics and Physics), students need to have gained a mathematics qualification prior to entering university-level study.

A considerable number of pre-higher education mathematics qualifications are available within the UK and, for those working within the higher education (HE) sector, it is not always clear what mathematics content, methods and processes students will have studied (or indeed can be expected to know and understand) as they commence their university-level programmes.

The Maths, Stats & OR Network, in conjunction with the Subject Centres for Bioscience, Engineering, Information and Computer Sciences, Materials and Physical Sciences, commissioned Mathematics in Education and Industry (MEI) to compile a mathematics guide. This outlines what students with given prior qualifications in mathematics are likely to know and be able to do and is written for those within the HE sector. Note that it does not include 'other' science qualifications which may include elements of mathematics and/or statistics in them.

## 1.2 Introduction and overview

This guide begins with a chapter setting the scene on pre-university qualifications and study. This includes an introduction to the main qualifications, a brief historical review of major developments and an overview of what and how entrants have studied prior to starting higher education.

The main content of the guide is encapsulated within a chapter on specific qualifications and the attributes of students who enter with them. Information about qualifications is given in short sections; if the user wishes to refer to a particular qualification it should be straightforward to identify the relevant section of the chapter.

A chapter is provided on useful sources of information. This is broken down into two parts, the first giving links to specific references raised in the previous chapter, and the second part on additional links to other documents (useful for gaining a more detailed understanding) and to relevant organisations (where information and updates can be found).

The guide concludes with appendices, including one on acronyms used in the guide and one which presents the statistics on the number of entrants to mathematics A Level over the last 20 years.

Overall this guide will give an overview of the key qualifications and offers links to further information that should aid the reader to gain an understanding of pre-university mathematics qualifications.

# 2. Setting the scene: pre-higher education qualifications and study

## 2.1 Introduction to the main qualifications

In March 2008 the Department for Children, Schools and Families published a consultation paper, 'Promoting achievement, valuing success: a strategy for 14-19 qualifications', which set out the government's intention to move towards a more streamlined and understandable qualifications framework for young people aged 14-19 in England. At the heart of this strategy are three main routes to higher education: apprenticeships, diplomas and general qualifications, including the General Certificate of Secondary Education (GCSE) and the General Certificate of Education, Advanced Level (GCE A Level). The GCSE is usually taken at 16 years of age and the GCE A Level after a further two years of study. For A Levels, students work at Advanced Subsidiary (AS) Level in their first year and at 'A2' Level in their second year. When the AS and A2 components are put together they form a full A Level qualification.

Apprenticeships combine paid work with on-the-job training, qualifications and progression. They do not include a requirement to take mathematics qualifications.

Diplomas offer a blend of classroom work and practical experience. They include a requirement to study functional mathematics at the appropriate level. All diploma lines of learning permit learners to include other mathematics qualifications. Further details are given in section 3.5.

General qualifications in mathematics provide the evidence of attainment in mathematics that is most likely to be presented to HE admissions tutors. This guide will clarify the content, style of assessment and probable learning outcomes that may be expected in a number of general qualifications in mathematics: these are GCSE, A Level and Free Standing Mathematics Qualifications (FSMQ).

## 2.2 Brief historical review of major developments

General qualifications in mathematics have developed in the context of widespread recent changes in expectations for learners. The government's objective of encouraging up to 50% of 18-30 year olds to attend higher education, for example, has influenced (albeit it gradually and without official decisions) the demands of both GCSE and A Level qualifications in mathematics.

The replacement of the General Certificate of Education, Ordinary Level (GCE O Level) by GCSE in 1988 may be seen as the start of a process by which these 'school leaving' qualifications could more closely reflect what the majority of 16 year olds know, understand and can do. Since that time, the substantial problem solving requirement of O Level Mathematics has been replaced by GCSE examinations that have required candidates to show capability in handling a broad range of basic mathematics questions.

Similarly, the introduction of subject cores for A Level examinations in 1983, the acknowledgement in 1996 that the AS standard should be pitched according to what is likely to be achieved a year before taking A Level and the rise of modular assessment at A Level since 1990 have all played significant parts in making A Level Mathematics examinations much more accessible than they were between 1951 and 1983. Until 1987 results were norm referenced so that in any subject 10% attained grade A, 15% B, 10% C, 15% D, 20% E and a further 20% were allowed to pass. This produced a bimodal distribution which did not match candidates' mathematical knowledge.

## 2.3 Where and how will entrants have studied pre-higher education?

It is important to be clear that those entering degree courses come from a wide range of backgrounds and bring with them a wide range of experiences. Two overarching factors relevant to this are *where* an entrant studied previously and *how*.

This information guide is made with particular reference to those entering onto a degree from a UK background (i.e. not overseas). With respect to the situation in England (Wales, Scotland and Northern Ireland will be dealt with in section 3.7) the major breakdown of categories of places of learning is in terms of age range, type and whether it is state-funded or independent (fee paying).

Over 90% of the secondary population attend state (government-funded) schools and, in the context of study prior to entry to higher education, establishments could include many different age ranges and have a varied focus. Age ranges for secondary study could involve 11-18, 14-19 or 16-19. The last of these could be small sixth forms attached to a school or they could be huge stand-alone Colleges of Further Education (FE) or Sixth Form Colleges. A learner may have been at the same place of study since the age of 11, or may have been at an establishment for only one or two years to complete their pre-higher education studies.

In terms of the independent sector it is widely expected that many of those attending such establishments will have been exposed to high quality tuition and learning resources and, although there is only a small proportion of the age cohort attending such establishments nationwide, most will go on to enter HE.

Having detailed the above, it is very difficult to definitively describe the way students will have been taught in all of these different establishments. Be it in the state or independent sector, some will have been in small classes whilst others will have been in large classes, some will have had well qualified teachers/lecturers, others non-qualified mathematics teachers. What is apparent, though, is that learners will enter HE with different experiences and respond to the relative changes that university-level study will bring in different ways. This is the case without even considering the specific subject knowledge which will be detailed in the next chapter.

# 3. Specific UK qualifications and attributes of students who enter higher education with them

This section describes the structure and content of specific UK mathematics qualifications and attempts to indicate the likely attributes of students who have taken them. However, the content of qualification specifications cannot be assumed to be an accurate measure of what students will actually know and understand when they start higher education. This will be influenced considerably by the nature of their mathematical learning experiences and by the grades they achieved.

Several universities have used diagnostic tests to determine the mathematical knowledge, understanding and fluency of new undergraduates, and how they relate to students' qualifications at the start of their HE courses.

GCSE and A Level qualifications are examined by three awarding bodies in England: Assessment and Qualifications Alliance (AQA), Edexcel and Oxford, Cambridge and RSA (OCR). They are regulated by the Office of the Qualifications and Examinations Regulator (Ofqual). The Qualifications and Curriculum Development Agency (QCDA) is the government agency responsible for developing these qualifications. GCSE and A level qualifications are also taken by students in Wales and Northern Ireland, though the

# Understanding the UK Mathematics Curriculum Pre-Higher Education

arrangements for administration are different. Scotland operates a separate system of examinations.

The large majority of students entering HE have taken GCSE and A Level qualifications, but several other qualifications are also used as routes into HE.

## 3.1 General Certificate of Secondary Education

### 3.1.1 Overview

Students in state schools have to follow the National Curriculum until age 16. GCSE Mathematics assesses the mathematics National Curriculum and is usually taken by students at the end of compulsory education (age 16). Some GCSEs follow a modular structure, with students taking some examinations in year 10 (age 15) and the rest in year 11 (age 16). Although the GCSE course is often thought of as a two year course, the GCSE work in mathematics builds directly on earlier work in mathematics and so the GCSE examinations test the mathematics that students have learnt throughout secondary school (11-16), and earlier. The content of GCSE Mathematics is the same for all awarding bodies, though it can be divided in different ways for modular courses. For GCSEs taken up to 2012, the content is specified by the 1999 National Curriculum, see (1).

Many students do not do any more mathematics after GCSE. Such students, who have not done any mathematics for two or three years before starting their degree courses, are likely to have limited recollection of GCSE content and techniques.

GCSE Mathematics is available at either Foundation Tier or Higher Tier. Grades C, D, E, F and G are available at Foundation Tier and grades A*, A, B and C are available at Higher Tier. Students who narrowly miss grade C at Higher Tier may be awarded grade D. Students who took GCSE Mathematics prior to 2008 may have taken it at Intermediate Tier, which allowed access to grades B, C, D and E. Grade C was not available at Foundation Tier until 2008. About 55% of students taking GCSE Mathematics achieve grade C or above.

Students entering GCSE Mathematics at Foundation Tier will not have studied as much mathematics as students taking Higher Tier. However, the grade boundary for grade C at Foundation Tier is higher than for C at Higher Tier, so Foundation Tier students with grade C have shown a good understanding of the mathematics which they have studied.

### 3.1.2 Subject knowledge and skills

Students who have not gone beyond the content of Foundation Tier GCSE will not have met some topics which students taking Higher Tier will have encountered. The list below covers the main topics *not* covered by Foundation Tier GCSE students:

- negative and fractional powers
- working with numbers in standard form (scientific notation)
- reverse percentage calculations
- working with quantities which vary in direct or inverse proportion
- solution of linear simultaneous equations by algebraic methods
- factorising quadratic expressions and solution of quadratic equations
- plotting graphs of cubic, reciprocal and exponential functions
- trigonometry
- calculation of length of arc and area of sector of a circle
- cumulative frequency diagrams, box plots and histograms
- moving averages
- tree diagrams and associated probability calculations.

Students who have been entered for Higher Tier Mathematics and achieved grade B or C will have an incomplete understanding of items from the list above and are likely to find algebra difficult.

GCSEs in Mathematics taken from summer 2012 will cover very similar content to the current ones but will put more emphasis on problem solving and functionality in mathematics.

### 3.1.3 International General Certificate of Secondary Education

The International General Certificate of Secondary Education (iGCSE) was originally designed for international schools but is now taken by students in some independent schools in the UK. iGCSE Mathematics is not available in state funded secondary schools as it is not fully aligned with the National Curriculum. The standard and content are similar to GCSE but students may have studied some additional topics, such as an introduction to calculus or matrices.

## 3.2 Advanced Subsidiary and Advanced Levels

### 3.2.1 Overview

The information below refers to Advanced Subsidiary and A Levels taken after the year 2000. Further changes are proposed for teaching from 2012 which may affect students taking A Level Mathematics in 2014.

AS Level Mathematics, Further Mathematics and Statistics each consist of three modules (also called 'units'). A Level in each of these subjects consists of six modules, which include the three AS modules. Students who have A Level will also have studied the AS content, but as they may not have requested the certification for the AS separately it might not appear on their certificate. The modules in these subjects are all of equal size.

The raw marks on each module are converted to Uniform Marks (UMS) to allow for slight differences in difficulty of examinations from year to year: the overall grade is decided by the total uniform mark gained on all modules. Students can resit individual modules to improve their marks. All modules are available in June with some also available in January.

The modules available in the MEI Mathematics and Further Mathematics A Levels are shown in Figure 1. (Note the MEI specification is administered through the Awarding Body OCR.) Similar structures apply for the other Mathematics and Further Mathematics A Levels.

AS Mathematics consists of the compulsory modules C1 and C2 and an applied module, which could be in mechanics, statistics or decision mathematics. A Level Mathematics has three further modules: the compulsory modules C3 and C4 and another applied module. The two applied modules in A Level Mathematics can be from the same area of applied mathematics or from different areas. The content of C1 and C2 together (AS) is nationally specified; likewise for C3 and C4 (A2). The content of applied modules varies between different exam awarding bodies. The national core can be found in the criteria for A Level Mathematics on the Ofqual website, see (2). This document also details what students who achieve grade A, C or E can typically do (this only gives a general idea as grades are based on total marks achieved rather than on these criteria, so strengths in some areas may balance out relative weaknesses in others). Students with the full A Level will have a broader knowledge of the AS core content than the A2 content because they have further developed their understanding in the second year.

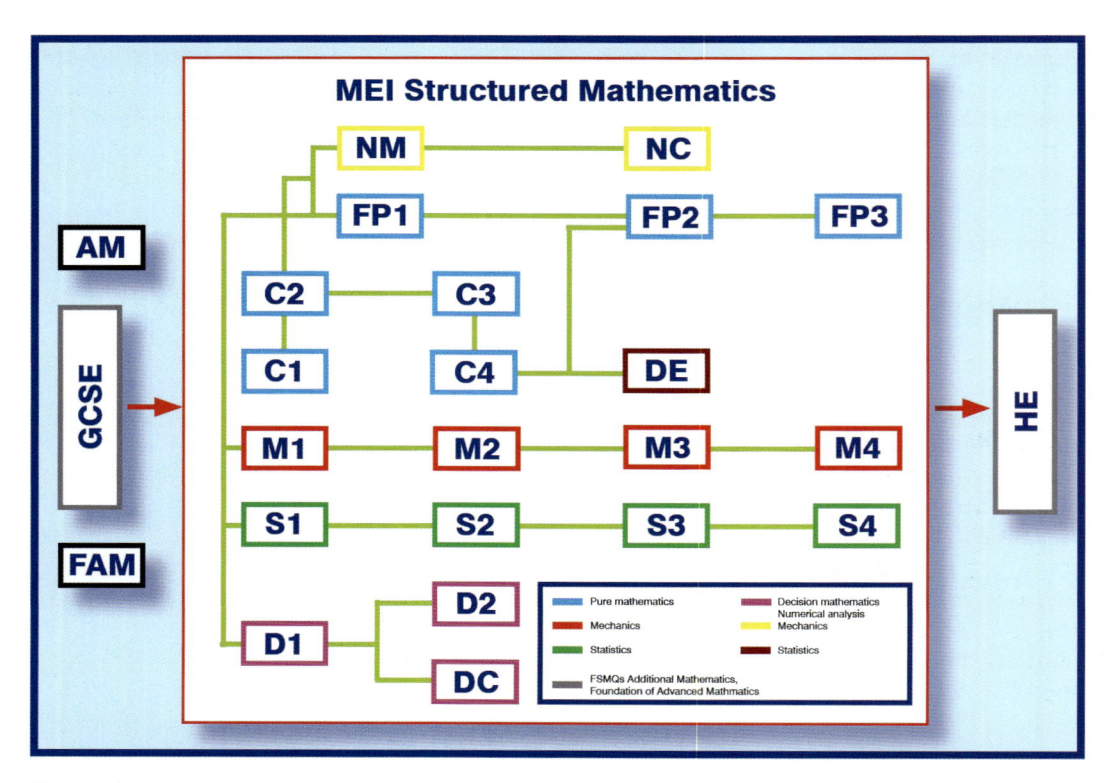

### Understanding the UK Mathematics Curriculum Pre-Higher Education

**Figure 1**

(Figure 1 notes – AM is Additional Mathematics, FAM is Foundations of Advanced Mathematics, NM is Numerical Methods, NC is Numerical Computation, FP is Further Pure Mathematics, C is Core Mathematics, DE is Differential Equations, M is Mechanics, S is Statistics, D is Decision Mathematics, DC is Decision Mathematics Computation.)

Typically, students complete AS Levels after one year. They may stop their study of mathematics at this point or go on to complete the full A Level in a further year. However, some schools enter students early for GCSE Mathematics and so they begin AS Mathematics in year 11 (age 16) and take three years to complete the full A Level. Other students take AS in year 13 (age 18) when their future plans are clearer.

Further Mathematics is only taken by students who are also taking Mathematics. They take three further modules for AS, including one compulsory module, Further Pure 1. To complete the A Level in Further Mathematics, students take at least one more pure module and two other modules. Students taking A Level Mathematics and A Level Further Mathematics will take 12 different modules and students taking A Level Mathematics and AS Level Further Mathematics will take 9 modules.

The optional modules in AS and A Level Further Mathematics can be drawn from either pure mathematics or applied mathematics. Applied modules are in suites for the three strands of applications: mechanics, statistics and decision mathematics. Mechanics 1 and Mechanics 2 could be taken in either A Level Mathematics or A Level Further Mathematics, but Mechanics 3 and 4 (and higher) are only available to students taking Further Mathematics. Similarly, for modules in statistics most awarding bodies only have two decision mathematics modules available.

For students who have taken both Mathematics and Further Mathematics AS and/or A Level, the Mathematics qualification consists of the compulsory core modules (C1 to 4) and a valid combination of applied modules. The remaining modules make up the Further Mathematics qualification. If there is more than one possible valid combination of applied modules to give A Level Mathematics, the combination of modules making up the separate AS or A Levels is automatically decided by the exam awarding body's computer in order to maximise the pair of grades students receive for Mathematics and Further Mathematics. The rules for aggregation and certification can be seen in (3).

A table of the numbers who have studied A Level Mathematics and Further Mathematics can be seen in appendix 5.2.

A small number of students take 15 modules to gain A Levels in Mathematics and Further Mathematics and AS Further Mathematics (Additional), and some take 18 modules for A Levels in Mathematics, Further Mathematics and Further Mathematics (Additional).

### *3.2.2 Subject knowledge and skills*

The vast majority of A Level students will be taught in schools and colleges and so will not be used to studying mathematics independently. Most A Level examination questions are structured. Past papers and specimen papers can be found on awarding bodies' websites and will give an idea of what students are expected to be able to do. Students who have taken both Mathematics and Further Mathematics will have greater fluency in the subject due to the greater amount of time they have spent on it. A document giving an overview of the content studied in mathematics A Levels can be seen in (4). This is also included in appendix 5.3.

Grades available at AS and A Level are A to E and U (where U is unclassified); grades achieved on individual modules are available to universities through UCAS, as well as the result for the whole qualification. From summer 2010 grade A* will be available for the full A Level (but not for AS). For A Level Mathematics, a total of 180 UMS marks (out of 200) will be needed on the two compulsory A2 modules (C3 and C4). For A Level Further Mathematics, a total of at least 270 (out of 300) is needed on the best three A2 modules. Students with A* will have shown the ability to work accurately under pressure.

A small number of students take AS or A Level Pure Mathematics. The A Level consists of the four compulsory core modules from A Level Mathematics together with two Further Mathematics Pure Modules. It cannot be taken with Mathematics or Further Mathematics AS or A Level.

Some students take AS or A Level Statistics. This is a separate qualification from Mathematics and Further Mathematics and the modules in it focus more on the use of statistics, whereas the statistics modules in the mathematics suite are more mathematical. The content of AS or A Level Statistics would be very useful background for students going on to study Business, Biology, Psychology or Social Sciences at higher education level.

Students who have completed their mathematical studies a year or more before starting higher education may need some support with revision to regain the fluency they had when they sat their examinations.

### **3.3 Advanced Extension Award and Sixth Term Examination Paper**

The Advanced Extension Award (AEA) is based on A Level Mathematics core content and is designed to challenge the most able students. It is offered by all of the awarding bodies but the examination paper is set by Edexcel. AEAs in other subjects exist but are being withdrawn - the Mathematics AEA will continue until at least 2012. The Mathematics AEA is assessed by a three hour paper of pure mathematics questions with no calculator allowed. Grades available are Distinction and Merit. Although candidates do not have to learn any additional content for AEA they do need to get used to a different style of question and to present clearly structured mathematical arguments.

## Understanding the UK Mathematics Curriculum Pre-Higher Education

The Sixth Term Examination Paper (STEP) is a university admissions test originally used only for entrance to Cambridge but it is now also used by some other universities. It is administered by the Cambridge Assessment examination board. There are three mathematics papers (I, II and III) and candidates usually take two of them. Paper III is based on A Level Further Mathematics and papers I and II on A Level Mathematics, but questions may include some content that is not in the A Level syllabus. However, candidates are not expected to learn extra content for the examination. Each paper has 13 questions: eight pure mathematics, three mechanics and two statistics. No calculator is allowed. Candidates are expected to answer six questions in three hours. Grades available are S (Outstanding), 1 (Very Good), 2 (Good), 3 (Satisfactory) and U (Unclassified). Usually a candidate will be awarded a grade 1 for a paper if they answer four (out of six) questions well.

Students who are successful in AEA or STEP will have a high level of ability to think for themselves, persist with a problem and present structured mathematical arguments.

### 3.4 Free Standing Mathematics Qualifications

FSMQs were first developed in the late 1990s. The initial motivation was to support vocational qualifications, e.g. General National Vocational Qualifications (GNVQ's), but it was also recognised that they could provide useful courses for other students as well. Uptake of the original FSMQs was not high and only AQA now offer them in their original form. They are tightly focused qualifications in applications of number, algebra, calculus, geometry, statistics or decision mathematics and they compensate for their narrow focus by requiring quite deep coverage.

OCR offers two FSMQs. These cover mathematics more broadly, but in less depth. *Foundations of Advanced Mathematics* (FAM) is a level 2 qualification that is designed to help bridge the gap between GCSE and AS Mathematics for students with a C/B grade in Mathematics GCSE. *Additional Mathematics* is a level 3 qualification aimed at able GCSE students and designed to be taken alongside GCSE. It is comparable in difficulty to AS Mathematics.

The AQA qualifications are likely to be used to support study of a range of courses in subjects other than mathematics. The OCR qualifications are more likely to be used to demonstrate achievement of a milestone in a learner's mathematical development.

All FSMQ qualifications are similar in size, rated at 60 guided learning hours (the same size as a single unit of an A Level that is divided into six units).

Level 3 units are awarded Universities and Colleges Admissions Service (UCAS) points.

The AQA FSMQs share a single assessment model. Candidates must produce a coursework portfolio worth 50% of the credit and sit a written examination for the remaining credit. The OCR FSMQs use slightly different assessment approaches, but both are assessed by written examination only.

Students who have achieved success in these qualifications are likely to share the broad capabilities of students achieving other mathematics qualifications at Levels 1, 2 and 3, see (5). However, students who have taken the AQA FSMQs will have demonstrated the ability to appreciate real world use and application of mathematics; they will also have engaged with completing a substantial coursework project. Students who have achieved success with OCR Additional Mathematics are likely to have shown an excellent grasp of basic advanced topics, which should be valued all the more highly if the qualification was taken pre-16. Students who have been successful in FAM will have studied a broader range of mathematics and are therefore more likely to be able to meet the demands of mathematics post-GCSE, particularly in algebra and trigonometry.

*3.4.1 AS Use of Mathematics*
Students who take level 3 FSMQs with AQA have an opportunity to use these as part of an AS qualification called AS Use of Mathematics. To achieve this requires two of the level 3 FSMQs, one of which must be *Working with algebraic and graphical techniques*, together with a terminal unit, *Applying Mathematics*.

A full A Level in Use of Mathematics is currently being piloted by AQA.

## 3.5 Diplomas
These qualifications for 14 to 19 year olds offer a blend of classroom work and practical experience. The level 1 diploma is equivalent to five GCSEs at grades D – G; the level 2 diploma is equivalent to seven GCSEs at grades A* - C; and the level 3 diploma is equivalent to 3.5 A Levels. Candidates taking level 1 and 2 diplomas must also take GCSE Mathematics. Note that there is also a slightly smaller version of the diploma (the progression diploma) and a slightly larger version (the extended diploma). More information about these may be seen at (6).

Diplomas consist of three curriculum parts:
- principal learning, the learning related to the title of the particular diploma
- generic learning, which consists of functional skills and a project
- additional specialist learning (ASL), which may be chosen freely from an extensive catalogue

From September 2009 a consortia of schools and colleges will offer diplomas in ten lines of learning, currently planned to increase to 17 by 2011. Of those published to date, only the level 3 Engineering Diploma has a mathematics unit as a compulsory part of its principal learning. This unit is called 'Mathematical Techniques and Applications for Engineers'. The unit is rated at 60 guided learning hours, the same size as a single unit of an A Level that is divided into six units. Many experts have suggested that the Mathematical Techniques and Applications for Engineers unit is significantly larger than a single unit of A Level Mathematics.

The level 3 diplomas include 360 guided learning hours of additional specialist learning. This is the size of an A Level, so learners who have completed level 3 diplomas could include A Level Mathematics as part of their diploma. For the level 3 Engineering Diploma, a purpose-designed mathematics ASL unit has been accredited and will be examined by OCR. Its title is 'Mathematics for Engineering' and it is rated at 180 guided learning hours, the same size as an AS qualification, but the unit does have a considerable amount of material, appearing to be larger than a typical AS in mathematics.

## 3.6 Other qualifications

### 3.6.1 International Baccalaureate
The International Baccalaureate (IB) Diploma Programme is recognised as a challenging two-year course for students aged 16-19. Around 190 schools in the UK now offer the qualification (as at 2009).

The course consists of a core (made up of three separate parts) and six subject groups, from which students select six subjects. Normally three subjects are studied at standard level (representing 150 teaching hours and broadly equivalent to AS Levels) and three are studied at higher level (representing 240 teaching hours and broadly equivalent to A Levels).

Four courses are available in mathematics: mathematics studies, mathematics or further mathematics at standard level, and mathematics at higher level.

These courses aim to enable students to develop mathematical knowledge, concepts and principles; develop logical, critical and creative thinking and employ and refine their powers of abstraction and generalisation.

All students who study the IB are required to study one of the mathematics qualifications. The full specification for mathematics at higher level can be seen in (7) and the full specification for further mathematics at standard level can be seen in (8).

### 3.6.2 Pre-U

The Pre-U is a recent qualification which began for first teaching in 2008, with first certification due to take place in summer 2010. It has a linear structure with examinations at the end of the two year course.

Students study at least three Principal Cambridge Pre-U subjects from a choice of 26, see (9). They also complete an Independent Research Report and a Global Perspectives portfolio. Mathematics and Further Mathematics are individual subjects in this list of 26. Full details of the Pre-U can be seen at (10).

It is too early to tell what a student will 'look' like having studied such a qualification and it is unknown how much take up there will be, but it is expected to be low. Those that do the Pre-U are almost certain to be from the independent sector.

### 3.6.3 Access courses

Access to Higher Education courses are specifically designed for people who left school without traditional qualifications but who would like to take a degree level course. Although there is a dedicated website where some information can be obtained, see (11), the content of courses varies. It may be advisable for institutions to review students entering with such qualifications on an individual basis to determine their prior mathematical experience.

### 3.6.4 Foundation courses

These courses are run by some universities to offer students unable to apply directly to a degree programme the opportunity to study in an area or subject of their interest and progress to a level where, on successful completion, they can enter onto an undergraduate course.

Although it is perhaps usual for a learner on such a course to progress onto a subsequent course at the institution at which they had studied their Foundation course, this need not necessarily be the case. However, these courses are institution specific and so it would again be advisable for institutions to review on a case-by-case basis what such a course entailed and thus what can be expected of students who have completed it.

### 3.7 Wales, Scotland and Northern Ireland

In Wales there is much overlap with England in the mixture of post-16 courses and the differing establishments available. However, most students sit examinations with the awarding body, the Welsh Joint Education Committee (WJEC). Some students educated in Wales are taught using English specifications and take examinations with the English awarding bodies. The regulatory body is the Department for Children, Education, Lifelong Learning and Skills (DCELLS).

Over the past five years (2004-2009) the number studying A Level Mathematics has steadily increased from circa 2500 to circa 3150. A Level Further Mathematics numbers are still comparatively low at 250 in 2009 (up from 140 in 2004).

# Understanding the UK Mathematics Curriculum Pre-Higher Education

There is also a Welsh Baccalaureate qualification taken by a small number of students, see (12).

In *Scotland*, at age 16 students take Standard Grade examinations (roughly equivalent to the English GCSEs), at age 17 they take Highers (equivalent to AS Levels) and then at 18 Advanced Highers (equivalent to A Levels). It is worth noting that although many students do stay on to study Advanced Highers, students going on to study at a Scottish university can gain entrance having studied only Highers.

Grades for Standard Grades are on a scale from 1 to 7, where 7 is the lowest and awarded if the course was completed but the examination not passed. There are 3 levels of difficulty of papers: Foundation (grades 5 and 6), General (grades 3 and 4) and Credit (grades 1 and 2). Each level has two papers, one which allows a calculator and one which does not. Both papers assess knowledge, understanding, reasoning and enquiring skills.

There are also two Intermediate courses, Intermediate 1 and Intermediate 2, which can be studied instead of Standard Grades, or taken alongside Highers. The Intermediate 1 Course is equivalent to the knowledge and skills developed in Standard Grade Mathematics at Foundation level and the Intermediate 2 Course is designed to be equivalent to the knowledge and skills developed in Standard Grade Mathematics at General level. Their structure is similar to that of Highers (i.e. a one-year course split into units). Further detail can be seen in (13).

There is only one examining body in Scotland, the Scottish Qualifications Authority (SQA). It makes the following remarks about an Advanced Higher in Mathematics:

> Advanced Higher Mathematics is designed to provide a broadened understanding of algebra, geometry and calculus for those candidates wishing to develop the experience they gained through the Higher Course.
>
> The Course will consolidate and extend the candidates' existing mathematical skills, knowledge and understanding in a way that recognises problem solving as an essential skill and that will allow them to integrate their knowledge of different areas of the subject.

There is also an Advanced Higher in Applied Mathematics, which enables students to study mathematics in real-life contexts and create and interpret mathematical models. However, the uptake for this is relatively low, with around 300 students studying it each year, see (14).

In *Northern Ireland* there are AS and A Levels in Mathematics and Further Mathematics. In 2009 approximately 2750 students studied A Level Mathematics, up from 2300 in 2004 but, again, relatively few study A Level Further Mathematics - 150 in 2009 (up from 140 in 2004). The examinations are provided by the Council for the Curriculum, Examinations and Assessments (CCEA), which is also the regulatory body.

# 4. Useful sources of information (to keep up-to-date with pre-higher education developments)

It is clear that there is much more to be said than can be included in a concise guide like this one. Links to relevant sources of information are given here. Most of these will consist of links to build upon information given earlier, but there are also some links to other material which would be useful for gaining a deeper understanding.

## 4.1 References made in this guide

(1) 1999 National Curriculum, see: http://curriculum.qcda.gov.uk/key-stages-3-and-4/subjects/key-stage-4/mathematics/index.aspx

(2) The national core for A Level Mathematics can be found in the criteria, see: http://www.ofqual.gov.uk/files/2002-12-gce-maths-subject-criteria.pdf

(3) The rules for A Level aggregation and certification can be seen at: http://www.jcq.org.uk/attachments/published/1047/GCE%20Maths%20Rules%20-%20guidance%20for%20centres.pdf

(4) Overview of content in mathematics A Levels: http://www.mei.org.uk/files/Maths_Alevel_Content_09.pdf

(5) National Qualifications Framework: http://www.qcda.gov.uk/libraryAssets/media/qca_05_2242_level_descriptors.pdf

(6) Extended diploma information can be seen at: http://yp.direct.gov.uk/diplomas

(7) Full specification for IB mathematics at higher level can be seen at: http://www.education.umd.edu/mathed/conference/vbook/math.sl.08.pdf

(8) Full specification for IB further mathematics at standard level can be seen at: http://www.education.umd.edu/mathed/conference/vbook/math.fut.06.pdf

(9) Information on the subjects available in the Pre-U can be seen at: http://www.cie.org.uk/qualifications/academic/uppersec/preu/subjects

(10) Full details of the Pre-U can be seen at: http://www.cie.org.uk/qualifications/academic/uppersec/preu

(11) Access courses website, see http://www.accesstohe.ac.uk/

(12) Welsh Baccalaureate, see http://www.wbq.org.uk/about-us

(13) Detail of Scottish Intermediate and Higher qualifications, see http://www.understandingstandards.org.uk/markers_ccc/mark_main.jsp?p_applic=CCC&p_service=Content.show&pContentID=532

(14) Scottish Mathematics External Assessment Reports: http://www.sqa.org.uk/sqa/2468.html

## 4.2 Additional references

### *4.2.1 Documents/information*

BTEC and OCR Nationals http://www.edexcel.com/quals/nat/Pages/default.aspx (and http://ocrnationals.com/level-3.php - level 3 is the Post-16 equivalent)

Cockcroft report (1982): http://www.dg.dial.pipex.com/documents/docs1/cockcroft.shtml (A report of the Committee of Inquiry into the teaching of Mathematics in Schools)

Evaluation of participation in A level mathematics: Final report, QCA (2007) http://www.ofqual.gov.uk/719.aspx

Guide to qualifications: http://www.direct.gov.uk/en/EducationAndLearning/QualificationsExplained/index.htm

Ideas from Mathematics Education - An Introduction for Mathematicians. Authored by Lara Alcock and Adrian Simpson. ISBN 978-0-9555914-3-3 http://www.mathstore.ac.uk/repository/AlcockSimpsonBook.pdf

International Baccalaureate http://www.ibo.org/who/ and http://www.ibo.org/diploma/curriculum/

iGCSE: http://www.cie.org.uk/qualifications/academic/middlesec/igcse/overview

Mathematical Association spreadsheet on the changes in school mathematics 1999-2016: http://www.m-a.org.uk/resources/TheChangingShapeOfTheCurriculum.xls

Mathematics: understanding the score, OFSTED (2008): http://www.ofsted.gov.uk/Ofsted-home/Publications-and-research/Documents-by-type/Thematic-reports/Mathematics-understanding-the-score/(language)/eng-GB

Maths careers website: www.mathscareers.org.uk

Report on mechanics at the transition from school to university: http://mathstore.ac.uk/repository/NewtonMechReportFinal.pdf

Scottish Standard Grades: http://www.scotland.gov.uk/library/documents/standard.htm

STEP: http://www.admissionstests.cambridgeassessment.org.uk/adt/step

Vocational qualifications: http://www.direct.gov.uk/en/EducationAndLearning/QualificationsExplained/DG_10039020

### *4.2.2 Organisations*

Two lists are presented in this sub-section. The first is of organisations which serve to provide additional information on mathematics and/or education (note that not all have previously been mentioned explicitly in this guide, but are included here for completeness); the second is of relevant Higher Education Academy (HEA) Subject Centres. Further information on strategies and resources aimed at supporting students with mathematics in the relevant discipline will be available from their websites.

- ACME www.acme-uk.org
- AQA http://web.aqa.org.uk/
- ATM http://www.atm.org.uk/
- CCEA http://www.rewardinglearning.org.uk/microsites/mathematics/gce/index.asp
- BIS http://www.dcsf.gov.uk/furthereducation/

**Understanding the UK Mathematics Curriculum Pre-Higher Education**

- DCSF http://www.dcsf.gov.uk/
- Edexcel http://www.edexcel.com/quals/gce/gce08/maths/Pages/default.aspx
- JCQ http://www.jcq.org.uk
- MA http://www.m-a.org.uk/jsp/index.jsp
- MEI www.mei.org.uk
- OCR http://www.ocr.org.uk/qualifications/subjects/mathematics/index.html
- OFQUAL http://www.ofqual.gov.uk
- Ofsted www.ofsted.gov.uk
- QCDA http://www.qcda.gov.uk
- SQA http://www.sqa.org.uk
- SCQF http://www.scqf.org.uk/

A complete list of HEA Subject Centres is available at:
http://www.heacademy.ac.uk/subjectcentres

- Bioscience (UK Centre) http://www.bioscience.heacademy.ac.uk/
- Engineering (Subject Centre) http://www.engsc.ac.uk/
- Information and Computer Sciences (Subject Centre) http://www.ics.heacademy.ac.uk
- Materials Education (UK Centre) http://www.materials.ac.uk
- Maths, Stats & OR (Network) http://www.mathstore.ac.uk/
- Physical Sciences (Centre) http://www.heacademy.ac.uk/physsci

# 5. Appendices

## 5.1 Acronyms used in this guide (including appendices)

| | |
|---|---|
| ACME | Advisory Committee on Mathematics Education |
| AEA | Advanced Extension Award |
| AS/A Level | Advanced Subsidiary/Advanced Level |
| ASL | Additional and Specialist Learning |
| AQA | Assessment and Qualifications Alliance |
| ATM | Association of Teachers of Mathematics |
| BIS | (Department for) Business, Innovation and Skills |
| CCEA | Council for the Curriculum, Examinations and Assessments |
| CSE | Certificate of Secondary Education |
| DCELLS | Department for Children, Education, Lifelong Learning and Skills (in Wales) |
| DCSF | Department for Children, Schools and Families |
| DfES | Department for Education and Skills |
| FAM | Foundations of Advanced Mathematics |
| FE | Further Education |
| FM | Further Mathematics |
| FSMQ | Free Standing Mathematics Qualifications |
| GAIM | Graded Assessment in Mathematics |
| GCSE | General Certificate of Secondary Education |
| GCE AS/A Level | General Certificate of Education Advanced Subsidiary/Advanced Level |
| GCE O Level | General Certificate of Education Ordinary Level |
| GNVQ | General National Vocational Qualification |
| HE | Higher Education |
| IB | International Baccalaureate |
| iGCSE | International GCSE |
| JCQ | Joint Council for Qualifications |
| MA | Mathematical Association |
| MEI | Mathematics in Education and Industry |
| OCR | Oxford Cambridge and RSA Examinations |
| OFQUAL | Office of the Qualifications and Examinations Regulator (in England) |
| OFSTED | Office for Standards in Education, Children's Services and Skills |
| QCDA | Qualifications and Curriculum Development Agency |
| SCQF | Scottish Credit and Qualifications Framework |
| SCUE | Standing Conference on University Entrance |
| SQA | Scottish Qualifications Authority |
| STEM | Science, Technology, Engineering and Mathematics |
| STEP | Sixth Term Examination Paper |
| UCAS | Universities and Colleges Admissions Service |
| UMS | Uniform Mark Scale |
| WJEC | Welsh Joint Education Committee |

## 5.2 A Level Mathematics Numbers 1989 – 2009 (Source JCQ)

| Year | Mathematics* entries (FM excl) | FM* entries | Total Mathematics entries (FM incl) | FM as % of Mathematics | Total A Level entries | Mathematics as % of total entries (FM incl) |
|---|---|---|---|---|---|---|
| 1989 | | | 84 744 | | 661 591 | 12.8 |
| 1990 | | | 79 747 | | 684 117 | 11.7 |
| 1991 | | | 74 972 | | 699 041 | 10.7 |
| 1992 | | | 72 384 | | 731 024 | 9.9 |
| 1993 | | | 66 340 | | 734 081 | 9.0 |
| 1994 | | | 64 919 | | 732 974 | 8.9 |
| 1995 | | | 62 188 | | 725 992 | 8.6 |
| 1996 | | | 67 442 | | 739 163 | 9.1 |
| 1997 | | | 68 880 | | 777 710 | 8.9 |
| 1998 | | | 70 554 | | 794 262 | 8.9 |
| 1999 | | | 69 945 | | 783 692 | 8.9 |
| 2000 | | | 67 036 | | 771 809 | 8.7 |
| 2001 | | | 66 247 | | 748 866 | 8.8 |
| 2002 | | | 53 940 | | 701 380 | 7.7 |
| 2003 | 50 602 | 5315 | 55 917 | 10.5 | 750 537 | 7.5 |
| 2004 | 52 788 | 5720 | 58 508 | 10.8 | 766 247 | 7.6 |
| 2005 | 52 897 | 5933 | 58 830 | 11.2 | 783 878 | 7.5 |
| 2006 | 55 982 | 7270 | 63 252 | 13.0 | 805 698 | 7.9 |
| 2007 | 60 093 | 7872 | 67 965 | 13.1 | 805 657 | 8.4 |
| 2008 | 64 593 | 9091 | 73 684 | 14.1 | 827 737 | 8.9 |
| 2009 | 72 475 | 10 473 | 82 948 | 14.5 | 846 977 | 9.8 |

* Note that prior to 2003 JCQ did not report Mathematics and Further Mathematics numbers separately.

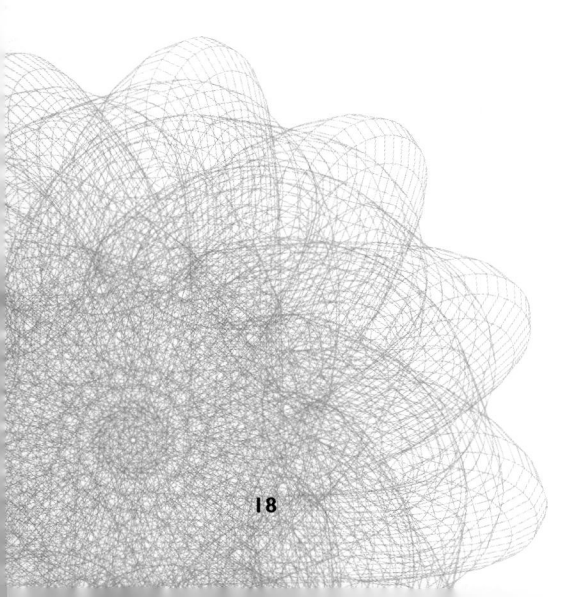

## 5.3 Overview of content in mathematics A Levels

### 5.3.1 What mathematics do students study in A level Mathematics courses?

Since the structure of A level Mathematics (and Further Mathematics) was changed in September 2004, students with only a single A level in Mathematics will have studied only two applied modules (in addition to the four core modules, Core 1 to Core 4, which cover the compulsory 'pure' content of the A level).

Please see http://www.meidistance.co.uk/pdf/revised_gce_maths_criteria_20040105.pdf for the official 'pure' core for AS/A level Mathematics, which must be covered by ALL AS/A level specifications.

Possible combinations of modules studied for A level Mathematics are:

| Core 1, Core 2, Core 3, Core 4 ||||||
| --- | --- | --- | --- | --- | --- |
| + one of the combinations of two applied modules shown below ||||||
| Statistics 1 | Mechanics 1 | Decision 1 | Statistics 1 | Decision 1 | Mechanics 1 |
| Mechanics 1 | Decision 1 | Statistics 1 | Statistics 2 | Decision 2 | Mechanics 2 |

There are no prescribed applied modules that are required to be studied, hence students could study any one of these combinations in order to gain an A level in Mathematics.

The following pages summarise the approximate content of the core modules in A Level Mathematics, in AS/A Level Further Mathematics and in the first two modules of each applied strand. However, there are differences between the content of such modules for the different A level specifications (and additionally a few other applied modules may be available from specific boards, e.g. Numerical Methods by MEI).

Those students who have studied an AS or A Level in Further Mathematics will have had the opportunity to study more applied mathematics modules than those with just the single A level Mathematics. This highlights the worth of the further mathematics qualification for those students who wish to study for mathematics-related degrees at university. Please see www.furthermaths.org.uk to find out more about Further Mathematics.

There is considerable scope for MEI to work with universities to help support the learning and teaching of undergraduates for both first year Mathematics courses and for mathematics service courses in mathematics-related degrees. For more information, please see www.mei.org.uk and navigate to 'University' under the 'What we do' menu.

If you have any queries, please contact:
**Charlie Stripp** (Charlie.Stripp@mei.org.uk)
or **Stephen Lee** (Stephen.Lee@mei.org.uk).

Understanding the UK Mathematics Curriculum Pre-Higher Education

# MECHANICS

**Motion Graphs**
- Displacement-time, distance-time, velocity time
- Interpreting the graphs
- Using differentiation and integration

**Newton's Laws Applied Along a line**
- Motion in a horizontal plane
- Motion in a vertical plane
- Pulleys
- Connected bodies

**Constant Acceleration and "SUVAT" Equations**
- Introduction to the variables
- Using the variables
- Standard properties
- Use in solving equations

**Vectors and Newton's Laws in 2 Dimensions**
- Resolving forces into components
- Motion on a slope (excluding and including friction)

**Projectiles**
- Finding the maximum height, range and path of a projectile

**Collisions**
- Coefficient of restitution
- Conservation of linear momentum
- Impulse
- Calculations involving a loss of energy

**Centre of Mass**
- Uniform bodies (symmetry)
- Composite bodies
- Centres of mass when suspended

**Equilibrium of a Rigid Body**
- Moment about a point
- Coplanar forces
- Toppling/sliding

**Variable Acceleration**
- Using differentiation in 1 and 2-D
- Using integration in 1 and 2-D

**Energy, Work and Power**
- Work done
- GPE, KE
- Conservation of energy
- Power (force does work)

**Uniform Motion in a Circle**
- Angular speed
- Acceleration
- Horizontal circle, conical pendulum

© MEI 2009

# STATISTICS

### Correlation and Regression
- Product moment correlation
- Spearman coefficient rank correlation
- Independent and dependant variables
- Least squares regression
- Scatter diagrams

### The Binomial Distribution and Probability
- Probability based on selecting or arranging
- Probability based on binomial distribution
- Expected value of a binomial distribution
- Expected frequencies from a series of trials

### Exploring Data
- Types of data
- Stem and Leaf diagrams
- Measures of central tendency and of spread
- Linear coding
- Skewness and outliers

### Normal Distribution
- Properties (including use of tables)
- Mean and variance
- Cumulative distribution function
- Continuous random variables (probability density function and mean/variance)
- As an approximation to binomial or Poisson distributions
- t-distribution

### Chi-squared
- Introduction
- Conditions

### Data Presentation
- Bar charts, pie charts
- Vertical line graphs, histograms
- Cumulative frequency

### Discrete Random Variables
- Expectation and variance of discrete random variables
- Formulae extensions $E(aX+b)$

### Probability
- Measuring, estimating and expectation
- Combined probability
- Two trials
- Conditional probability
- Simple applications of laws

### Hypothesis Testing
- Establishing the null and alternate hypothesis
- Conducting the test and interpreting the results
- Use of the binomial or normal distribution
- Type 1 and type 2 errors

### Poisson Distribution
- Properties (including use of tables)
- Mean and variance
- Use as an approximation to binomial distribution

### Sampling/Estimation
- Randomness in choosing
- Sample means and standard errors
- Unbiased estimates of population means
- Use of central limit theorem
- Confidence intervals

© MEI 2009

Understanding the UK Mathematics Curriculum Pre-Higher Education

# DECISION

**Graphs**
- Graphs

**Networks**
- Prim
- Kruskal
- Dijkstra
- TSP
- Route inspection
- Network flows

**Critical Path Analysis**
- Activity networks
- Cascade charts

**Game Theory**
- Game theory
- Using simplex

**Optimisation**
- Matchings
- Hungarian algorithm
- Transportation problem
- Dynamic programming

**Algorithms**
- Communicating
- Sorting
- Packing
- Efficiency and complexity

**Discrete Random Variables**
- LP graphical
- LP simplex
- Two stage simplex

**Simulation**
- Monte Carlo methods
- Uniformly distributed random variables
- Non-uniformly distributed random variables
- Simulation models

**Logic and Boolean Algebra**
- Logical propositions and truth tables
- Laws of Boolean algebra
- Combinatorial circuits and switching circuits

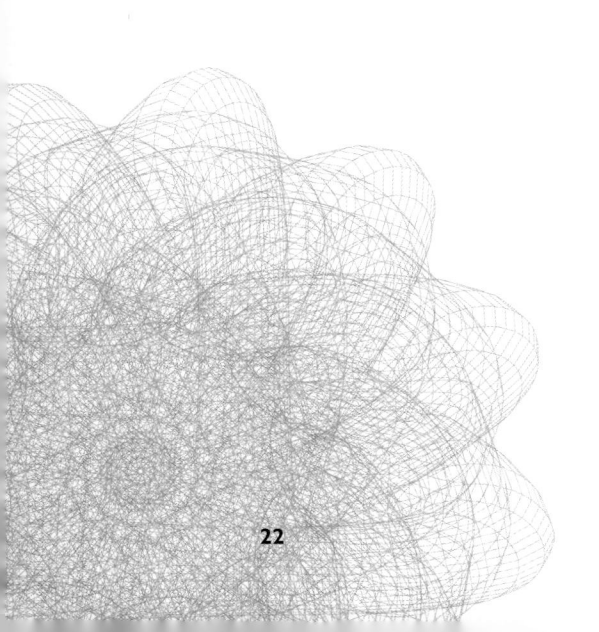

© MEI 2009

## 5.4 Important dates for Mathematics (authored by Roger Porkess)

| Year | Up to 16 | 16-19 |
|---|---|---|
| Earlier | School certificate | Higher school certificate |
| 1950 | Start of O Level | |
| 1950 | Start of O Level | |
| 1951 | | Start of A Level |
| 1962 | First CSE examinations | |
| 1969 | | Proposal by SCUE to replace A Levels with Q and F Levels (later rejected) |
| 1970s | Introduction of 16+ examinations combining CSE and O Level (running alongside both) | |
| 1973 | | Proposal by the Schools Council to replace A Levels with N and F Levels (later rejected) |
| 1973-4 | Raising of the school leaving age to 16 | |
| 1978 | Calculators allowed in O Level | |
| 1982 | Cockcroft Report | |
| 1983 | | First A Level common cores for mathematics and sciences |
| 1986 | | New syllabuses based on core |
| 1986-8 | O Level and CSE replaced by GCSE | |
| 1987 | | Norm referenced grading of A Levels replaced by criterion referenced grading |
| 1988 | Coursework in GCSE Mathematics (not compulsory at first) | Higginson Report recommends wider sixth form curriculum (instantly rejected) |
| 1989 | Introduction of the National Curriculum | |
| 1990 | | First teaching of first modular A Level syllabus (MEI Structured Mathematics) |
| 1992 | First teaching of revised GCSEs to assess the National Curriculum | |
| 1993 | Externally set and teacher marked National Curriculum tests at ages 7, 11 and 14 | Second A Level common cores for mathematics and sciences

First core for AS Mathematics |
| 1994 | Teacher boycott of marking National Curriculum tests | |
| 1994 | | First teaching of new syllabuses based on new cores

Almost all syllabuses are modular |
| 1996 | | The Dearing Review on 16-19 Education |
| 1997 | New rules require exactly 3 tiers of entry. Several syllabuses (e.g. GAIM and MEI are lost) | New A Level syllabuses developed but put on hold following the general election |

## Understanding the UK Mathematics Curriculum Pre-Higher Education

| Year | Up to 16 | 16-19 |
|---|---|---|
| 1999 | National Numeracy Strategy (primary schools) | New common cores at AS and A Level, now known as subject cores |
| 2000 | | First teaching of new syllabuses to conform with Curriculum 2000 |
| 2000 | | Work starts on an MEI pilot programme to foster Further Mathematics |
| 2001 | Key Stage 3 National Strategy Framework for teaching mathematics to Years 7, 8 and 9<br><br>Introduction of data handling coursework at GCSE | Very poor results at AS Level<br><br>Marked drop in retention rate for full A Level |
| 2002 | | Large reduction in numbers taking A Level mathematics |
| 2003 | | New subject cores for AS and A Level Mathematics |
| 2004 | The Smith Report: *Making Mathematics Count* ||
| 2004 | The Tomlinson Report on 14-19 Curriculum and Qualifications Reform (mostly rejected) ||
| 2004 | | First teaching of new mathematics syllabuses, now known as specifications, introduced to overcome the problems caused by Curriculum 2000 |
| 2005 | | DfES rolls out the MEI Further Mathematics programme as the Further Mathematics Network (still run by MEI) |
| 2006 | First teaching of GCSE syllabuses that have changed from 3-tier to 2-tier | Uptake starts to rise under the new specifications |
| 2007 | First teaching of GCSE Mathematics with no coursework | |
| 2009 | Key Stage 3 National Curriculum tests discontinued | Uptake is the highest since 1989 |
| 2009 | | The Further Mathematics Network is replaced by the Further Mathematics Support Programme (still run by MEI) |
| 2010 | New GCSE syllabuses to start | |
| 2010 | Pilot of twin GCSEs to start | |